Usability Engineering and Human Factors for Medical Devices

An Introduction

Bernadette White

ISBN: 9798680239209
© 2022

Contents

Usability Engineering and Human Factors for Medical Devices 1
Introduction 5
Key Terms 7
Human Skills and Abilities 10
Vision 11
Visual acuity 11
Visual acuity can be represented with five measurements which include: 11
Visual Threshold 11
Focusing 11
Perception 11
Color 12
Audition & Speech 12
Information Processing 12
Memory 13
Usability Engineering Process 14
Inputs into Risk Identification 17
Preliminary Evaluations 18
Critical Task Identification and Categorization 18
Identification of Known Use-Related Problems 18
Analytical Approaches to Identifying Critical Tasks 19
Task Analysis 19
Heuristic Analysis (hands-on, informative) 21
Expert Review 21
Empirical Approaches to Identifying Critical Tasks 21
Contextual Inquiry 22
Interviews 22
Usability Engineering Process Overview 23
Use Related Risk Analysis 32
Product Realization and Usability Engineering 33
Product Realization per ISO 13485 Requirements: 33

pg. 3

ISO 13485 & Product Realization 34
Planning of Product Realization / Design and Development Planning 34
Usability Engineering Plan 36
Product Realization Process and Risk Management 36
Formative Studies and Evaluation 36
Efficiency 39
Safety 40
Usability Engineering File 40
Use Scenario 41
Environmental Factors and Usability 41
Use Specification 42
User Interface 43
User Needs and Requirements 44
User Interface Evaluation Plan 45
Summative Testing 46
Summative Evaluation Protocol 46
Post-Marketing Surveillance 47
European Regulations- Usability and MDR 50
Design Controls and Usability Engineering 51
Information Supplied and Usability 58
Useful Definitions 71

Introduction

Recent changes in regulation (e.g. MDR 2017/745) have increased the focus on usability requirements for medical devices. With stronger references to risks associated with use error and foreseeable misuse now requiring manufacturers to respond to foreseeable misuse.

If medical devices are designed and developed without applying a usability engineering or human factors engineering, their use can be non-intuitive, difficult to learn, difficult to complete tasks and difficult to use. Furthermore, as technology and medical devices offer more innovative solutions, patients can now be tasked with using medical devices or administering their treatments, therefore usability becomes more important.

While the goal of design should aim to provide medical devices that are inherently safe, as with most medical devices, residual risks remain once a product is designed, manufactured and validated clinically. Use errors or Usability errors contribute to those potential risk scenarios where medical device usability is an issue for the user. With the increasing abundance of medical devices is the observation, treatment and monitoring of patients, use errors must be assessed for medical devices and reduced to an acceptable level. In contrast to safety inherited by design, the least and often the last protective measures are warnings or contraindications provided on labelling or instructions for use.

The strength of applying usability engineering principles (aka human factors engineering) medical devices is that use errors can be identified and mitigated through design and engineering practices, early-on in the product development and product realization cycle. Medical devices designed and developed devoid of usability engineering, are less intuitive, difficult to use and require focus and attention to learn to use them effectively. In addition, usability is a growing requirement from a regulatory perspective.

The Usability Engineering Process and the subsequent activities should be planned in order to provide a roadmap of deliverables and ensure the requirements are fulfilled. The execution of studies must be executed, documented and approved by appropriate personnel with adequate training, education or experience.

The Usability Engineering Effort for a particular medical device can be estimated based on factors such as:

- Complexity or size of the User Interface, including readability for an IFU
- Complexity of the Use Specification (environment, user etc.)
- Severity of the harm associated with the use of the medical device

In general terms, the following usability questions can be used to understand some of the characteristics of the device.

- Is it easy to learn how to use the device?
- Do users remember how to use the medical device after periods of non-use (days, months)?
- How efficiently can the device be used?
- Is the device designed and manufactured in such a way that prevents users from making errors or allow the user recover from their use errors?
- Is the device appropriate for the user profile, taking into account their abilities?

While safety and performance are the principle concern for medical device manufacturers, usability engineering can be applied looking at non safety related tasks that users may complete. This can be beneficial for the overall user experience and benefit the manufacturer from a commercial point of view.

User needs

Establishing the needs of the user early on in the product development lifecycle can help inform the usability objectives of a medical device. This voice of customer or stakeholder research helps ensure that the product has the right design inputs identified making the development process as effective as possible.

Redesign

During the development of a device, the device or a prototype should be evaluated to determine if user interfaces are correctly specified and designed. Usability testing (formative) is extremely valuable in providing this feedback. If unanticipated use errors are discovered, there is time to address them if it is not close to the design transfer activities. Redesign of design changes resulting from evaluations and observations may be the subject of design change control.

Environmental Factors and Use Error

According to the purpose and intended use of a device, there may be environmental factors that can distract the user or if under challenging circumstances, making it more difficult for the user to not follow the instructions or not operate the device properly. Outdoor settings, hot and cold environments, ambulances or noisy environments are use scenarios that prevent specific challenges.

During usability evaluation, real world settings should be considered in order to obtain the most useful information that can be used as feedback for design teams. Establishing a Use Specification and testing of the User interface are important in this regard.

Quality by Design

Inherent or built in quality that supports safe use of a medical device is the desired approach. When a manufacturer uses the principles of usability engineering and meets the requirements of usability standards, they position themselves for greater success, by identifying usability issues and addressing them by with design solutions. While instructions for use and labelling provide important information and help to mitigate against some risks, the focus should be eliminating or reducing the risks at the device level rather than through documentation.

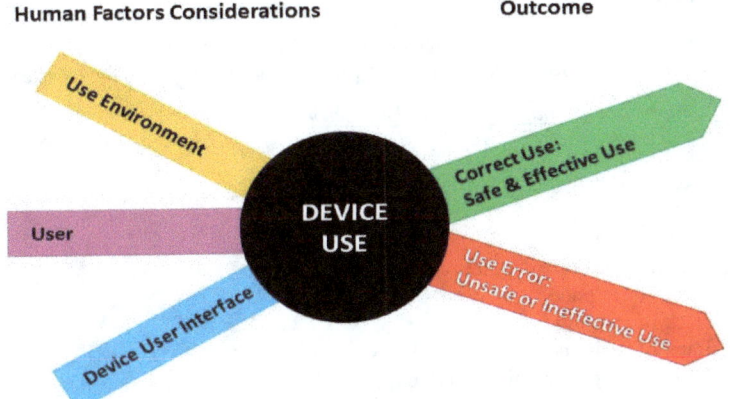

Source: *Guidance for Industry and Food and Drug Administration Staff: Applying Human Factors and Usability Engineering to Medical Devices, FDA.*

Key Terms

Use Error

The formal definition of Use Error is when *"User action or lack of user action (omission) while using the medical device that leads to a different result than that intended by the legal manufacturer or expected by the user"*. It is the "inability of the user to complete a task". It also should be noted that Users may or may not be aware that a use error has occurred.

Use errors can be caused be tasks not been followed per instructions, used in the incorrect environment or not for the intended use or indications. Unexpected physiological responses of the patient is not by itself considered use error. Nor is a malfunction of a medical device that causes an unexpected result. Use error occurs if the user is not able to complete a task. Use errors can result from poor understanding between the principles of operation, the characteristics of the user, user interface, task, or use environment.

A use error occurs at the "action" stage of the interaction with the device. Therefore, at the stage of perception (e.g. misreading a display) or at the stage of cognition (e.g. misinterpreting a number) are not considered use errors. Errors in perception and errors in cognition are classed as contributing factors to or causes of use errors. Hence, a use error (incorrect action or lack of action) can be caused by a misreading or by a misinterpretation of the medical device output, but the use error manifests itself only when an incorrect (aka erroneous action or lack of action takes place.

List of Potential Use Errors
- Error of commission, user performs incorrect action
- Error of omission, incorrectly omitting (failing to complete) a necessary action

List of Potential Factors that can lead to Use Errors
- Environmental distractions
- Excessive workload or stress
- Tiredness
- Lack of attention
- Working fast or too quickly
- Over confidence
- Lack of training
- Lack of experience
- Lack of language abilities (fluency)
- Interruptions

Correct Use

Correct use is when the user successfully uses the medical device and no use errors are encountered.

Normal use

During Normal use, a use error can occur while the user attempts to use the medical device in accordance with its instructions for use. Normal use encompasses all foreseeable user actions when a user is operating a medical device according to the manufacturer's intended use. Normal use is simply what is expected from a user under normal conditions of use, which includes actions that are either correct or in error.

Normal use is differentiated from intended use. Intended use addresses the medical purpose while normal use includes the medical purpose but also the storage, transportation, maintenance (if applicable) and so on.

Abnormal use is defined separately.

Abnormal use

Abnormal use includes use of the device with exceptional disregard for the intended use and instructions or disregarding the contraindications.

Reckless use

Reckless use involves situations where the user is not concerned with the potential danger of their actions.

Sabotage

Normally understood as related to a formative or summative evaluation where post testing the user admits they made a conscious decision to ignore instructions or not to complete an action.

Identification of Use Errors

The process of conducting Usability (Engineering) studies plays a key role in identifying scenarios where reasonably foreseeable misuse occurs.

Use Error is defined as a "user action or lack of user action while using the medical device that leads to a different result than that intended by the manufacturer or expected by the user"

Use Difficulty

Use Difficulties include repeated attempts to complete a task such as:
- ✓ hesitating,
- ✓ excessive "exploring" of the interface
- ✓ unexpectedly referring to the labeling information

Close Call

When a user makes a Use Error but then takes an action to "recover" and prevent the harm from occurring. Close calls may highlight problems with the design of the user interface.

Success

Usability testing or usability engineering studies can be performed during the development of a new product. It acts as a verification that a device is designed appropriately and can identify scenarios or conditions that users could present a use error or usability risk to the patient or user.

As defined above, Use Error is defined as a "user action or lack of user action while using the medical device that leads to a different result than that intended by the manufacturer or expected by the user" Technical report, ISO TR/24971:2020. This covers the following errors:

-the inability of the user to complete a task.

-Use errors resulting from a mismatch between the characteristics of the user, user interface, task, or use environment. Users may be aware or unaware that a use error has occurred.

Exception (to a use error):

- An unexpected physiological response of the patient is not by itself considered use error.

-A malfunction of a medical device

Critical task

A user task which, if performed incorrectly or not performed at all, would or could cause serious harm to the patient or user, where harm is defined to include compromised medical care. A task usually has a specific goal in mind.

Effectiveness

A measure of the accuracy and completeness in which a user achieves specified goal or outcome.

UOUP, User Interface of Unknown Provenance

User interface or a part of user interface previously developed where adequate records of the usability engineering process of recognized standards are not available.

Human Skills and Abilities

Design engineers and Human factors engineers must appreciate the abilities and limitations humans in how we see, sense, perceive, process and respond to environments when completing actions or tasks such as the use of a medical device. Not only is an understanding of the senses important but also how various senses depend on one another and how they interact. For example, if information is viewed on a HMI, the user may need to process the information in order to make a decision and then follow up with action by selecting a button or operation on the medical device. If the data is complex, the information processing by the user can take some time and hence influences the response time or the time until action is initiated.

Vision

Visual acuity

Visual acuity can be represented with five measurements which include:

- minimum distinguishable, (how easily detail can be detected)
- minimum perceptible, (detection of spot)
- minimum separable, (detection of a gap between parts)
- stereoscopic acuity, (detection of depth for 3D object)
- vernier (lateral) acuity, (detection of displacement on a line from another)

Physical and environmental factors can affect visual acuity including

- Level and kind of illumination,

- Object Size and color,
- Background,
- Is the object stationary or moving?
- Is the viewer stationary or moving?

<u>Visual Threshold</u>

This is the minimum light level in which an object can be identified visually. Vision depends on light been focused on the retina. The amount of light is a factor in the visual sensitivity.

<u>Focusing</u>

A percentage of the population experience myopia (shortsightedness) and hyperopia (long-sightedness) and stigmatism. These focusing complaints can be addressed with vision correction and also using larger symbols and increased text size on labelling, documentation and user interfaces (display screen). Age is a factor in the focusing abilities of humans and this should be accounted for in design inputs and definition of the user interface specification.

<u>Perception</u>

Visual perception is the ability to perceive surroundings using the available light that enters our eyes. The visual perception is of particular interest in relation to graphical user interfaces (GUIs). Visual perception occurs in the brain's cerebral cortex where the electrochemical signals get there by traveling through the optic nerve and the thalamus. The colors size and brightness of objects contribute to the physiological process

Color

The main situational or environmental factors that influences a person's color vision is the brightness and the nature of the light with Hue and saturation are also playing a role. There are various clinical conditions of color deficiency also. 'Green blindness' or 'red blindness' while rare perhaps form the most prevalent types of defective color vision. A very rare portion of the population will exhibit monochromatism- a loss of color completely accompanied with visual poor acuity.

Audition & Speech

Hearing abilities of individuals can depend on medical conditions and other factors such as age. Therefore, if a medical device makes audible sounds or alarms during the course of its operation, the age profile and disabilities of the user may need to be understood.

The aging process results in a decline in hearing sensitivity for human beings. The degree of hearing loss results from various physiological processes and effects from lifestyle.

- Changes in physiology of inner ear
- Reduction in cognitive processing
- Accumulative effects of loud sounds

Information Processing

Information processing in the human brain relies on sensory data and inputs. While many sources of information may be presented at once, the human mind tends to focus their attention on one source of sensory data at a time. The performance of individuals to complete tasks which require information processing is subject to a number of influencing factors.
- Time/duration of the sustained attention period
- Accuracy of the tasks act hand
- Speed at which the tasks must be completed
- Secondary tasks or other sources of distraction

Memory

> Sensory Memory
>
> Short-term Memory
>
> Long-term Memory

The three types of memory include sensory, short-term and long-term memory.

Sensory memory is a form of short-term memory received through the five senses (sight, hearing, taste, touch, and smell.) A sensory response as a result of a stimulus results in the memory been briefly stored. This type of memory can last only for a fraction of a second. When a person is riding a bicycle, the encounter many forms of stimuli- wind, air temperature, vibration, sounds and sights. This information does not need to be retained longer than a mere fractions of a second.

Short-term memory has a limited capacity to store information and data and is noted to have fast retrieval time also exhibits a fast loss. It has been often compared to a buffering process in a computer system. The information is required for imminent or short term use and therefore needs to be readily available. As short-term memory is a daily and minute by minute requirement it is also referred to as working memory. From a usability perspective, consider the use of an automated blood pressure monitor by a physician. Once the measurement is taken and displayed on the screen, the blood pressure reading or value is now subject to short-term memory. However, the user interface should consider the time the display holds the value on-screen in order to facilitate note taking or other actions by the user of physician.

Long-term memory has a greater capacity for information storage but is subject to slower retrieval time. Long term memory can be reinforced by practicing a skill and hence tends to see a slow rate of decay.

Usability Engineering Process

Each manufacturer should establish a Usability Engineering Process within their organization. This is necessary to embed usability engineering in product development and through the lifecycle of medical device produces. The core purpose of any Usability Engineering Process is to make the medical device user interaction safer. This requires usability problems and use errors to be identified and mitigated by ensuring all known or foreseeable hazard-related use scenarios are addressed.

The strengths of adapting a process approach to engineering is well regarded and evident by the principles of ISO 9001 and ISO 13485. The Usability process must not only be established but needs to be implemented (for each product), maintained and known to be effective during the lifecycle of a product. For a process to properly function, there are some essential activities needed. However, for a process to be effective it depends on its level of integration with other related processes and data generated from processes. For example, the usability process depends on post marketing surveillance data been generated and allowing usability issues to be identified.

Fundamental Principles of a Process

- Define the inputs and outputs from the processes
- Determine the interaction of processes (how to feed into one another)
- Determine the criteria and methods to ensure effective processes
- Determine the resources needed and availability
- Identify risks and prevent, reduce or if they cannot be eliminated enhance controls
- Improve processes and implemented changes

The usability engineering cycle emphasizes the iterative nature of the development process. The usability engineering process for medical devices is principally focused on providing safety for the patient, user and others related to usability. Applying a usability engineering process mitigates risk caused by usability problems associated with correct use and use errors under normal use. As with most engineering activities within regulated medical device industry, other processes within a quality management system, support activities such as a Risk management process. A risk management process must be established, maintained and be effective in documenting and identifying hazards and hazardous situations associated with a medical device. The usability effort draws upon the risk management procedures, policy and process as a whole. As the outputs of usability need to be evaluated and reviewed to determine any impact on the benefit-risk profile, required controls and overall risk acceptability.

Factors that determine the Usability Engineering effort

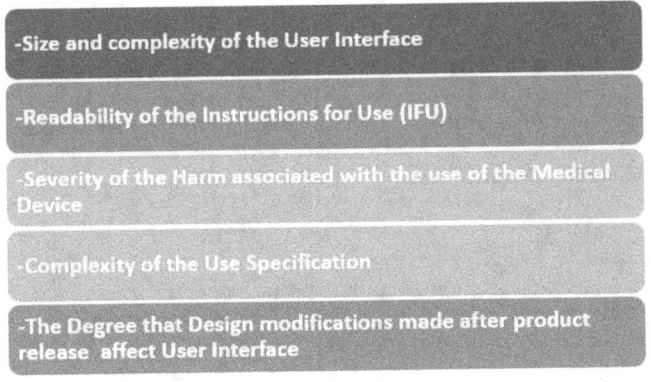

- Size and complexity of the User Interface
- Readability of the Instructions for Use (IFU)
- Severity of the Harm associated with the use of the Medical Device
- Complexity of the Use Specification
- The Degree that Design modifications made after product release affect User Interface

Estimating and evaluating the associated risks, controlling those risks, and monitoring how effective those controls is a continual process that occurs throughout the life-cycle of a product.

Risk management is a continuous process carried out over the product life cycle

A usability process assists a manufacturer in the analysis, specification, design and evaluation of the usability of a medical device. The risk management process been interrelated to the usability takes the usability of the device into account. Use errors are minimized by:
      a) discovering hazards and hazardous situations related to the user interface
      b) designing and implementing measures to control the risks related to the user interface
      c) evaluating the risk control measures.

All known or foreseeable hazard-related use scenarios are addressed prior to selecting those hazard-related use scenarios which are then used in preparing the user interface evaluation plan. It is also useful to be mindful of the meaning of a safe medical device. Safety can be understood as freedom from unacceptable risk. Unacceptable risk can arise from use error, which can lead to exposure to direct physical hazards or loss or degradation of clinical functionality.

Formative Evaluation
Formative studies or evaluation is performed early-on and throughout the develop process using simulations and early working prototypes that explore if general usability principles and intended use can be performed. These formative studies can inform and help develop suitable user interfaces and identify is any design changes are required.

Summative Evaluation

Summative Evaluations are completed during the design validation stage of a project. As a validation activity, it should have formal acceptance criteria. Summative testing is typically more detailed and specific than formative testing

Inputs into Risk Identification

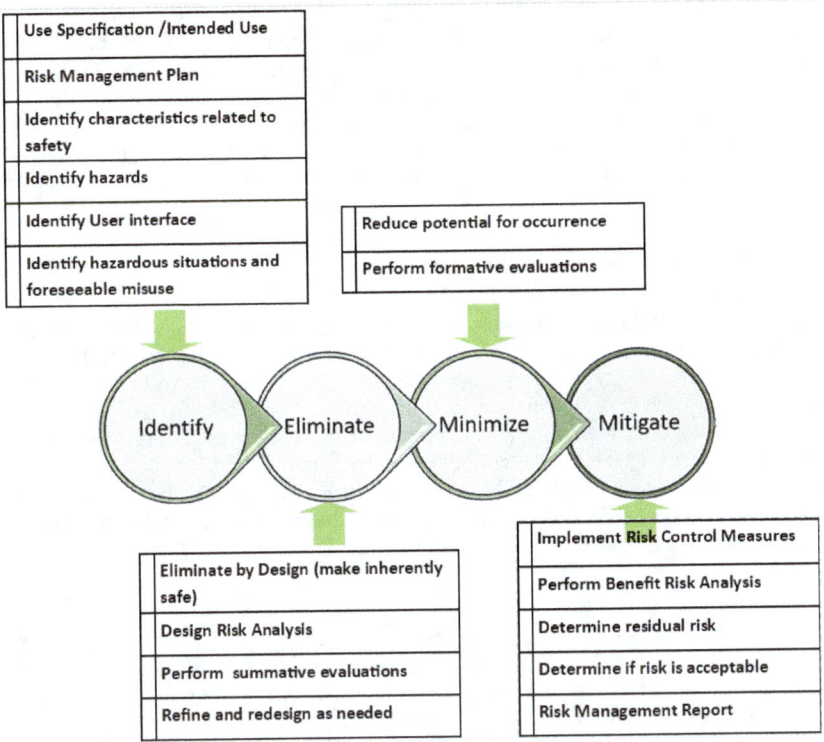

The purpose of any usability engineering process is to identify use errors and eliminate or minimize the use associated risks.

Preliminary Evaluations

Preliminary analyses and evaluations are performed to identify user tasks, user interface components and use issues early in the design process. These analyses help focus the HFE/UE processes on the user interface design as it is being developed so it can be optimized with respect to safe and effective use.

One of the most important outcomes of these analyses is comprehensive identification and categorization of user tasks, leading to a list of critical tasks. Human factors and usability engineering offer a variety of methods for studying the interactions between devices and their users. Analytical methods and empirical methods can be useful for identifying use-related hazards and hazardous situations. These techniques are discussed separately; however, they are interdependent and should be employed in complementary ways. The results of these analyses and evaluations should be used to inform your risk management efforts and development of the protocol for the human factors validation test.

Critical Task Identification and Categorization

An essential goal of the preliminary analysis and evaluation process is to identify critical tasks that users should perform correctly for use of the medical device to be safe and effective. You should categorize the user tasks based on the severity of the potential harm that could result from use errors, as identified in the risk analysis. The purpose is to identify the tasks that, if performed incorrectly or not performed at all, would or could cause serious harm. These are the critical tasks. Risk analysis approaches, such as failure modes effects analysis (FMEA) and fault tree analysis (FTA) can be helpful tools for this purpose. All risks associated with the warnings, cautions and contraindications in the labeling should be included in the risk assessment.

Reasonably foreseeable misuse (including device use by unintended but foreseeable users) should be evaluated to the extent possible, and the labeling should include specific warnings describing that use and the potential consequences. Abnormal use is generally not controllable through application of HFE/UE processes. The list of critical tasks is dynamic and will change as the device design evolves and the preliminary analysis and evaluation process continues. As user interactions with the user interface become better understood, additional critical tasks will likely be identified and be added to the list. The final list of critical tasks is used to structure the human factors validation test to ensure it focuses on the tasks that relate to device use safety and effectiveness.

Identification of Known Use-Related Problems

When developing a new device, it is useful to identify use-related problems (if any) that have occurred with devices that are similar to the one under development with regard to use, the user interface or user interactions. When these types of problems are found, they should be considered during the design of the new device's user interface. These devices might have been made by the same manufacturer or by other manufacturers. Sources of information on use-related problems include customer complaint files, and the knowledge of training and sales staff familiar with use-related problems. Information can also be obtained from previous HFE/UE studies conducted, for example, on earlier versions of the device being developed or on similar existing devices. Other sources of information on known use-related hazards are current device users, journal articles, proceedings of professional meetings, newsletters, and relevant internet sites, such as:

- FDA's Manufacturer and User Facility Device Experience (MAUDE) database
- FDA's MedSun: Medical Product Safety Network
- CDRH Medical Device Recalls
- FDA Safety Communications
- ECRI's Medical Device Safety Reports
- The Institute of Safe Medical Practices (ISMP's) Medication Safety Alert Newsletters

Analytical Approaches to Identifying Critical Tasks

A number of analytical approaches can be used to review and assessment of user interactions with devices. These approaches are most helpful for design development when applied early in the process. The results include identification of hazardous situations, i.e. specific tasks or use scenarios including user-device interactions involving use errors that could cause harm. The results are used to inform the formative evaluation and human factors validation testing that follow. Analytical approaches for identifying use-related hazards and hazardous situations include analysis of the expected needs of users of the new device, analysis of available information about the use of similar devices, and employment of one or more analytical methods such as task analysis and heuristic and expert analyses.

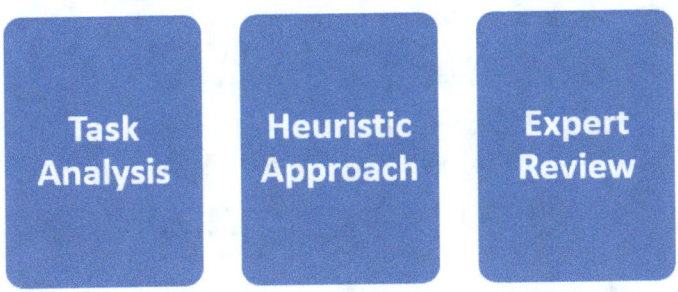

Task Analysis

Task analysis or critical tasks analysis techniques systematically break down the device use process into discrete sequences of tasks. The tasks are then analyzed to identify the user interface components involved, the use errors associated with each step and potential use errors. The use of task analysis can be used to help answer the following questions:

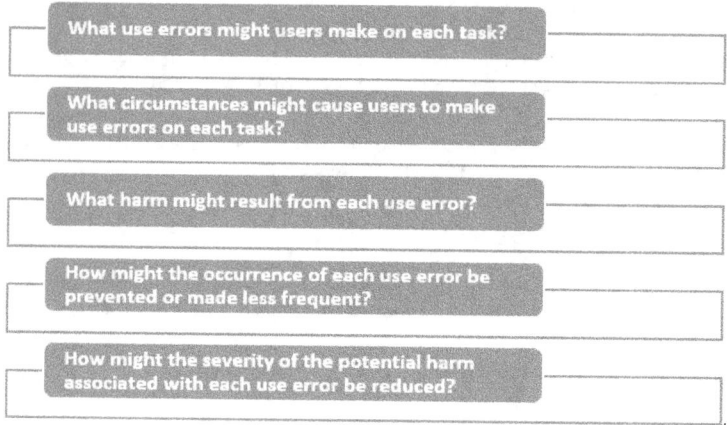

- What use errors might users make on each task?
- What circumstances might cause users to make use errors on each task?
- What harm might result from each use error?
- How might the occurrence of each use error be prevented or made less frequent?
- How might the severity of the potential harm associated with each use error be reduced?

Task analysis techniques can be used to study how users would likely perform each task and potential use error modes can be identified for each of the tasks. For each user interaction, the user actions can be identified by understanding the perceptual inputs, cognitive processing, and physical actions involved in performing the step.

For example, perceptual information could be difficult or impossible to notice or detect and then as a cognitive component they could be difficult to interpret or could be misinterpreted; additional cognitive tasks could be confusing or complicated or inconsistent with the user's past experiences; and physical actions could be incorrect, inappropriately timed, or impossible to accomplish. Each of these use error modes should be analyzed to identify the potential consequences of the errors and the potential resulting harm. The following questions should be considered:

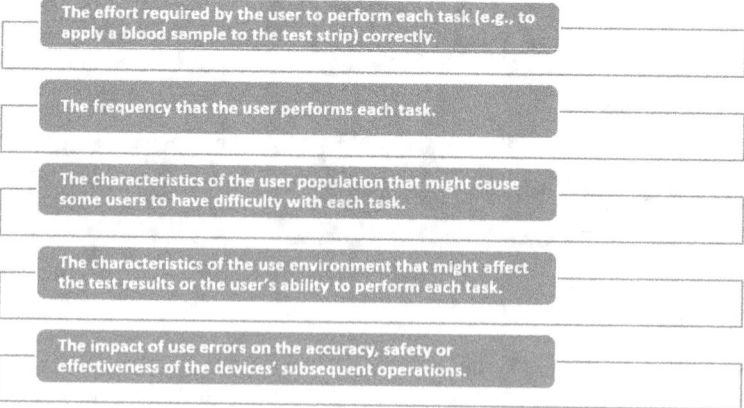

- The effort required by the user to perform each task (e.g., to apply a blood sample to the test strip) correctly.
- The frequency that the user performs each task.
- The characteristics of the user population that might cause some users to have difficulty with each task.
- The characteristics of the use environment that might affect the test results or the user's ability to perform each task.
- The impact of use errors on the accuracy, safety or effectiveness of the devices' subsequent operations.

Heuristic Analysis (hands-on, informative)

Heuristic analysis is a process in which analysts (usually HFE/UE specialists) evaluate a device's user interface against user interface design principles, rules or "heuristic" guidelines. The object is to evaluate the user interface overall, and identify possible weaknesses in the design, especially when use error could lead to harm. Heuristic analyses include careful consideration of accepted concepts for design of the user interface. A variety of heuristics are available and you should take care to select the one or ones that are most appropriate for your specific application.

Expert Review

Expert reviews rely on clinical experts or human factors experts to analyze device use, identify problems, and make recommendations for addressing them. The difference between expert review and heuristic analysis is that expert review relies more heavily on assessment done by individuals with expertise in a specific area based on their personal experiences and opinions.

The success of the expert review depends on the expert's knowledge and understanding of the device technology, its use, clinical applications, and characteristics of the intended users, as well as the expert's ability to predict actual device use. Reviews conducted by multiple experts, either independently or as a group, are likely to identify a higher number of potential use problems.

Empirical Approaches to Identifying Critical Tasks

Empirical approaches to identifying potential use-related hazards and hazardous situations derive data from users' experiences interacting with the device or device prototypes or mock-ups. They provide additional information to inform the product development process beyond what is possible using analytical approaches.

Empirical approaches include methods such as contextual inquiry, interview techniques and simulated-use testing. To obtain valid data, it is important in such studies for the testing to include participants who are representative of the intended users. It is also important for facilitators to be impartial and to strive not to influence the behavior or responses of the participants.

Contextual Inquiry

Contextual inquiry involves observing representatives of the intended users interacting with a currently marketed device (similar to the device being developed) as they normally would and in an actual use environment.

The objective is to understand how design of the user interface affects the safety and effectiveness of its use, which aspects of the design are acceptable, and which should be designed differently. In addition to observing, this process can include asking users questions while they use the device or interviewing them afterward. Users could be asked what they were doing and why they used the device the way they did.

This process can help with understanding the users' perspectives on difficult or potentially unsafe interactions, effects of the actual use environment, and various issues related to workload and typical workflow.

Interviews

Individual and group interviews (the latter are sometimes called "focus groups") generate qualitative information regarding the perceptions, opinions, beliefs and attitudes of individual or groups of device users and patients. In the interviews, users can be asked to describe their experiences with existing devices, specific problems they had while using them, and provide their perspectives on the way a new device should be designed. Interviews can focus on topics of particular interest and explore specific issues in depth.

They should be structured to cover all relevant topics but allow for unscripted discussion when the interviewee's responses require clarification or raise new questions. Individual interviews allow the interviewer to understand the perspectives of individuals who, for example, might represent specific categories of users or understand particular aspects of device use or applications. Individual interviews can also make it easier for people to discuss issues that they might not be comfortable discussing in a group. Group interviews offer the advantage of providing individuals with the opportunity to interact with other people as they discuss topics.

Usability Engineering Process Overview

Prepare Use Specification

- Include intended use and indications
- intended user profile and patient population
- Intendened body part of tissue device interacts with
- Use environment
- Operating Principle

Identify user interace characteristics and potential use errors

- What are the primary operating functions
- How does the user interface address safety and use errors

Identify known of foreseeable harzards and hazardous situations

- Identify hazards that could arise that affect patients
- Use Risk analysis tools to identify use errors

Identify Harzard-related Use scenarios

- List hazardous situations with hazardous Use scenarios
- Record tasks for use scenario and the severity of harms (E.g. Use-related risk assessment)

Create User Interface Specifiation

- Consider content of Use specification, known or foreseeable use errors.
- User interface specifiction should identify testable technical requirements and if training/documentation is required

Create User Interface Evaluation Plan

- Identify the method and scope of formative and summative studies, user profiles, test environment

Perform User Interface Formative Evaluation

- Via usability engineering methods, complete formative evalutions to assess the design. Revise is new user errors or hazards are identified

Perform Summative Evaluation

- Improve User interface if required
- If not practicle provide rationale
- determine residual risk

Usability and Risk Management

A manufacturers approach to Risk Management forms an essential part of regulatory requirements but also the identification of risks and how they are treated. In this respect, when risk management is applied diligently, it makes for a safer medical device where performance issues and potential hazards can be addressed in the design of the device in a preventative manner. Risks and hazards associated with medical devices can be divided into two separate categories.

1) Device failures that are unrelated to usability. For example;

- mechanical
- physical
- chemical
- functional failures
- biochemical
- Sterility

2) Use-Related Hazards by as a result of action on lack of action by the user resulting in a harm.

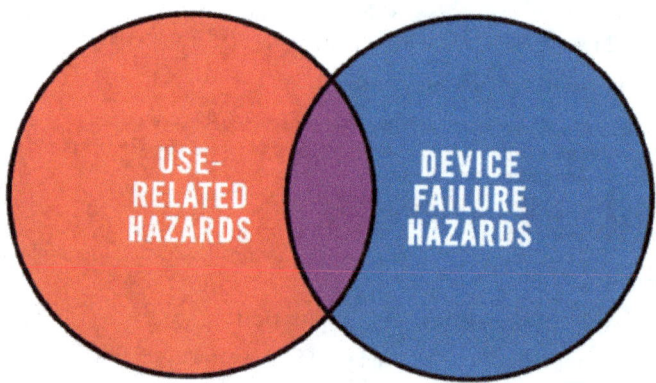

Source: *Guidance for Industry and Food and Drug Administration Staff: Applying Human Factors and Usability Engineering to Medical Devices, FDA.*

User errors are different to component failures of failures in functionality. It is known that estimating the probability of use errors occurring is a challenge, therefore focusing on the severity of the potential have is more important that the combination of severity and probability of the error occurring.

IEC 62366-1 Medical Devices, Application of usability engineering to medical devices is referenced both in ISO 14971 (Medical Devices- Application of Risk Management) and ISO/TR 24971 (Guidance of the application of risk management). While Risk Management and Usability Engineering are separate processes, they both supplement and overlap in their intent.

Usability engineering can determine if a certain misuse is reasonably foreseeable or not. The completion of a usability test can identify when users may frequently use the medical device in a manner that does not follow the manufacturer's instructions for use. Causes of misuse identified in testing can be a result of several factors such as:

- Poor working culture
- inadequate perception of risk
- limited knowledge of the consequences
- operating procedures /instructions are not clear

Therefore, risk management provides the tools and a decision-making process for estimating and evaluation these usability risks and determining if risk reduction is required or if the potential hazards or hazardous situations can be deemed acceptable risk Foreseeable hazards and risks can be a result of the design, functionality, layout or complexity of the user interface.

- ISO 14971 requires that risks associated with each of the identified hazardous situations be estimated and evaluated
- The manufacturer must establish a risk acceptability policy and criteria
- If a risk is not acceptable using the manufacturer's risk acceptability, appropriate risk control measure(s) that reduce the risk(s) to an acceptable level
- These controls must be implemented and verified as effective in reducing the risk to a pre-determined acceptable level.

Risk Management of Use Errors

This section covers techniques that identify and control use error in devices:

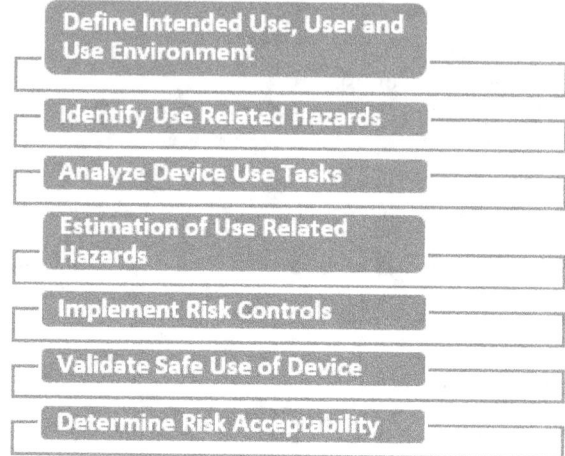

The above elements are described in the following section.

Define Intended Use, User and Use Environment

- The Intended use must be defined for the product early on as this represents the beginning of risk management. The intended use also informs the required design inputs
- The intended use is a short description of the medical device and its purpose
- The user must be defined bearing in mind the use scenarios also. A user may be a lay person, physician, nurse or other healthcare professional. The definition of the user provides perspective on what level or training and technical knowledge the user may have.
- The Use environment may require specific design inputs to protect products from influences of the operational environment. More specifically for usability, depending on the use scenario, the manufacturer and design may need to take into account factors such as stress, noise, outdoor use. This knowledge and context can then inform the user interface design to make interaction more intuitive for the user and minimise use-error.

Identify Use Related Hazards

- Identification of use-related hazards should commence early during the development of a product. The identification of use related hazards can be achieved via task analysis, input from stakeholders and users and formative and summative studies.
- Analysis of similar devices can provide a source of potential use errors and their occurrence levels

Analyze Device Use Tasks
- This technique helps to identify the user tasks in detail. Once critical tasks are identified, corresponding user requirements and user interface requirements can be realized.
- For each user requirement, potential failure modes from a usability perspective.

Estimation of Use Related Hazards
- Estimating the severity and occurrence of use errors can assist in organising and prioritising use-related risks and hazards.
- Failure modes effects and analysis is a risk assessment technique that ranks risks in terms of severity, occurrence and detection. The scores of the severity x occurrence x detection provide a Risk Priority Number, RPN. Other methods of risk assessment include Fault Tree Analysis

Implement Risk Controls
- Identified hazards are preferably eliminated using design features. This may require design modification to a current prototype. These changes should be agreed by a cross functional team. In addition, it should be highlighted that engineering or design features that may control use-errors can themselves present other risks. E.g. mechanical or software failure. However, not all use-errors can be eliminated by design changes and residual risks remain.
- In certain circumstances a change to the intended use or redefining the target or user population can be an option to the manufacturer if other protective measures or mitigations are not feasible.
- Training requirements on the correct use of the medical device should be considered. Training is susceptible to been less effective over time.
- Warning statements, symbols and labelling with they provide a degree of protective measures, their effectiveness depends on whether or not the user notices, reads, understands and follows the meaning of such controls.

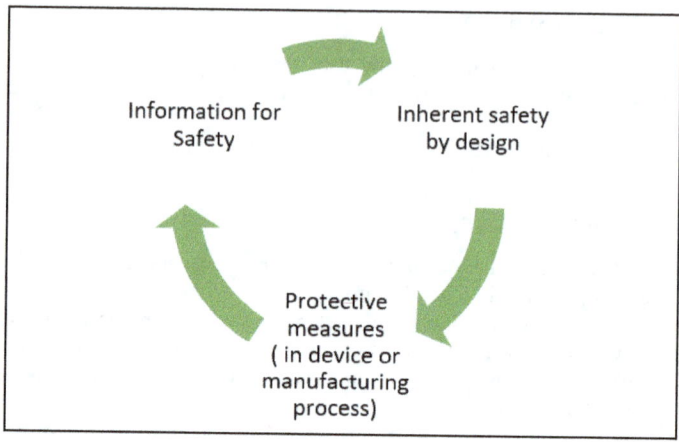

Validate Safe Use of Device

- Device safety can be best demonstrated by validation where the validation takes into account the intended use been applied by the user in a realistic environment.
- The validation should use an appropriate sample of users that has statistical rationale
- The final (or final stage) device should be used in conducting the validation in order to provide reliable results

Determine Risk Acceptability

- After applying a usability engineering process where use specification, user profiles, Use Related Risk analysis; have been completed iteratively through the design cycle, and formative and summative evaluations are concluded and where appropriate, actions taken, the residual risks for use-error should be very low. A determination on whether the remaining risk is acceptable must be made in the wider risk management context and in keeping with the manufacturers risk acceptability policies.

Risk Management Plan

A risk management plan describes the scope of the risk management activities (methods) along with the responsibilities and authorities, verification activities, verification methods, the production and post-production information to be collected and reviewed for the medical device and the criteria for risk acceptability.

Risk Acceptability Matrix

	S0 None	S1 Negligible	S2 Minor	S3 Serious	S4 Critical	S5 Catastrophic
P5 Frequent	Acceptable	Acceptable	Acceptable	Unacceptable	Unacceptable	Unacceptable
P4 Probable	Acceptable	Acceptable	Acceptable	Unacceptable	Unacceptable	Unacceptable
P3 Occasional	Acceptable	Acceptable	Acceptable	Acceptable	Unacceptable	Unacceptable
P2 Remote	Acceptable	Acceptable	Acceptable	Acceptable	Unacceptable	Unacceptable
P1 Rarely	Acceptable	Acceptable	Acceptable	Acceptable	Acceptable	Unacceptable

The risk management plan must be reviewed and updated throughout the lifecycle of the medical device as new information becomes available. Any changes to the risk management plan must also be recorded in the risk management file. An important factor to consider with regard to the level of detail should be commensurate with the level of risk associated with the medical device.

Identification of hazards from use errors

The usability testing or studies can highlight if issues occur when the device is used by the patient- for example, do people use the medical device in a way that it is not intended to be used or not in accordance with the instructions for use.

Hazards from reasonably foreseeable misuse

Some hazards and hazardous situations may be a result of reasonably foreseeable misuse. Engineering usability studies can also help identify and confirm reasonably foreseeable misuse scenarios.

Identification of hazards relating to Usability

The below series of questions are accompanied with an explanation and factors that should be considered when identifying hazards.

1) **Is successful application of the medical device dependent on the usability of the user interface?**

Depending on the function and intended use of the device most devices shall require certain critical tasks to be done in order to use the device. The usability of the user interface can determine if use-errors are likely to occur. A well-defined, developed and evaluated user interface minimizes greatly the risk of use errors.

2) **Can the user interface design features contribute to use error?**

Factors that should be considered include: control and indicators, symbols used, ergonomic features, physical design and layout, hierarchy of operation, menus for software-driven medical devices, visibility of warnings, audibility of alarms, standardisation of colour coding.

3) **Is the medical device used in an environment where distractions can cause use error?**

   Factors that should be considered include:
   — the consequence of use error;
   — whether the distractions are commonplace;
   — whether the user can be disturbed by an infrequent distraction;
   — whether repetitive stress can reduce the user's awareness or attention.

4) Does the medical device have connecting parts or accessories?

   Factors that should be considered include the possibility of wrong connections, similarity to other products' connections, connection force, feedback on connection integrity, and over- and under tightening.

5) **Does the medical device have a control interface?**

   Factors that should be considered include spacing, coding, grouping, mapping, modes of feedback, blunders, slips, control differentiation, visibility, direction of activation or change, whether the controls are continuous or discrete, and the reversibility of settings or actions.

6) **Does the medical device display information?**

   Factors that should be considered include visibility in various environments, orientation, the visual
   capabilities of the user, populations and perspectives, clarity of the presented information, units, colour coding, and the accessibility of critical information.

7) **Is the medical device controlled by a menu?**

   Factors that should be considered include complexity and number of layers, awareness of state, location of settings, navigation method, number of steps per action, sequence clarity and memorization problems, and importance of control function relative to its accessibility and the impact of deviating from specified operating procedures.

8) **Is the successful use of the medical device dependent on a user's knowledge, skills and abilities?**

   Factors that should be considered include:
   — the (intended) users, their mental and physical abilities, skill and training;
   — the use environment, ergonomic aspects, installation requirements;
   — the personal characteristics of intended users that can affect their ability to successfully interact with the medical device.

9) **Will the medical device be used by persons with specific needs?**

   Factors that should be considered include:
   — users with special characteristics, such as disabled persons, the elderly and children, who might need assistance by another person to enable the use of a medical device;
   — users having wide-ranging skill levels and differing cultural backgrounds and expectations that could lead to differences in what is considered appropriate application of the medical device.

10) **Can the user interface be used to initiate unauthorised actions?**

    Factors that should be considered include whether the user interface allows the user to enter an operation mode with restricted access (e.g. for maintenance or special use), which increases the possibility of use error and thereby the associated risks, and whether the user becomes aware of having entered such operation mode.

11) **Does the medical device include an alarm system?**

    Factors that should be considered are the risk of false alarms, missing alarms, disconnected alarm systems, unreliable remote alarm systems, and the user's ability of understanding how the alarm system works.

12) **In what ways might the medical device be misused (deliberately or not)?**

Factors that should be considered are incorrect use of connectors, disabling safety features or alarms, neglect of manufacturer's recommended maintenance, unauthorized access to the medical device or to medical device functions.

13) **Is the medical device intended to be mobile or portable?**

Factors that should be considered are the need for grips, handles, wheels or brakes, and the need for mechanical stability and durability.

Use Related Risk Analysis

Use Related Risk Analysis (URRA) is a risk analysis format and methodology that is used to identify and analyze hazards and harms arising from user errors and unintended use of a product/device. The Use Related Risk Analysis (URRA) is part of the risk management process for the medical device and forms part of the risk management file.

The initial use risk assessment should be performed as patient or user needs and inputs are developed. In the event new risks are identified that require mitigation follow up evaluations should be completed to identify protective measures.

After a product is launched and commercialized, the URRA still is available element of risk management. Post-Production Surveillance information collected will be evaluated for Use Errors and the Use Related Risk Analysis is updated as necessary.

IEC 62366-1 Medical devices — Part 1: Application of usability engineering to medical devices

IEC 62366-1 is an international standard that provides a framework to medical device manufacturers to analyse, specify, develop and evaluate the usability of a medical device as it relates to safety in a process-based approach. As described previously, the usability engineering process works to assist the manufacturer in assessing and mitigating risks associated with correct use and use errors during normal use.

IEC 62366 specifies a framework and process for a manufacturer to analyse, specify, develop and evaluate the usability of a medical device as it relates to safety. The systematic evaluation of devices for usability (aka human factors) is intended to identify and minimise use errors and in turn reduce use-associated risks. Once use errors are identified it allows the manufacturer to respond and mitigate as necessary.

The usability engineering (human factors engineering) process as prescribed in IEC 62366 allows a manufacturer to assess and mitigate risks associated with correct use and use errors, i.e., normal use. it can be used to identify but does not assess or mitigate risks associated with abnormal use.

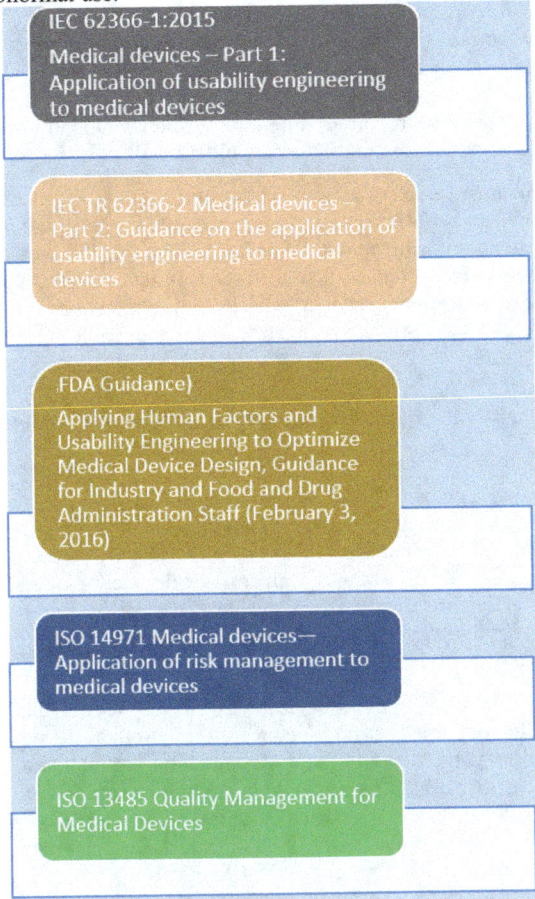

Product Realization and Usability Engineering

Product Realization per ISO 13485 Requirements:

- Planning of product realization 7.1
- Customer-related processes 7.2
- Design and development 7.3
- Purchasing 7.4
- Production and service provision 7.5
- Control of monitoring and measuring equipment 7.6

ISO 13485 & Product Realization

Product realization per ISO 13485 is the process of planning product development and introduction but also the subsequent steps that are meaningful in the success of the product introduction. Planning should be initiated early-on in the design stage and should include timelines, resources required, intended markets. One of the most important aspects of planning is gaining the correct stakeholder needs.

Product realization must establish customer requirements and document the design and development efforts. ISO 13485 also has requirements around purchasing, production, service product and monitoring and measuring equipment. Product realization can be defined as a collection of processes that delivers a product or service to the customer. There is an 'opt out' mechanism that where an organization can exclude specific requirements, in cases where product realization is not applicable.

Planning of Product Realization / Design and Development Planning

It is the manufacturer's responsibility to establish and maintain plans that describe or reference the design and development activities and define responsibilities for implementation. The plans should identify and describe the interaction with different groups or activities that are part of the design and development process. The maintenance of plans to reflect an accurate state as the design and development progresses is also a key factor. The design and development planning is intended to be prospective in nature. It allows risks to be identified earlier and promotes timely delivery of projects.

Product realization refers to the product development process that forms part of a product's life cycle, beginning with a conception, through its development, design iterations, verification, validation and to final completion. Product realization for medical devices is defined in ISO 13485 as a mandatory requirement for a quality management system (QMS). Product realization when adopted by a company and when applied consistently can provide benefits to the customer and the manufacturing company. It supports regulatory compliance and successful applications but can also better deliver products that meet customer and company requirements.

7.1 PLANNING OF PRODUCT REALIZATION

Planning is the first step in product realization and provides stakeholder alignment and a roadmap to success. Planning your product realization requirements should include:

- Marketing requirements that may specify product size, features, technology, colour, cost
- Targeted Quality requirements for the product
- Timelines for delivery of project milestones
- Product Intended Use and User Profiles
- Information on product verification and validation

- Applicable standards to be met e.g. Usability, sterility, product specific standards

7.2 CUSTOMER-RELATED PROCESSES

This requirement mandates organizations to fulfill all customer needs, whether those are explicitly stated by clients, necessary for an intended use, or required by regulatory standards. It also covers the best practices for communication of customer feedback. According to the ISO 13485 standard, a customer-related process extends further to delivery, post-delivery activities, and user training requirements.

7.3 DESIGN AND DEVELOPMENT
As designs go through different iterations and modifications, they must continuously reviewed against the inputs, with the outputs maintained as accurate and reflective of the design. It covers the following:

- Intended use of the medical device
- Requirements of regulation, quality and standards
- Product validation and verification
- Design transfer
- Design History requirements
- Control of Design changes

7.4 PURCHASING

The overall performance and quality of the supply chain management must be considered in the context of purchasing, selecting vendors during product realization.

7.5 PRODUCTION AND SERVICE PROVISION

Per ISO 13485, the production process should be planned, executed, monitored, and controlled so that it meets your specifications. Service provision may not be applicable to all products.

7.6 CONTROL OF MONITORING AND MEASURING EQUIPMENT

The control and monitoring of product is supported by measurement and inspection equipment. Ensuring this equipment is controlled, monitored and maintained to be reliable and consist is important in delivering safe and effective products.

Activities such as calibration should be completed in accordance with documented procedures, external standards and where nonconformances occur, a process for controlling product and addressing issues is available.

Usability Engineering Plan

A Usability Engineering Plan (UEP) should be established during the planning stage of product design and development. Sometimes the plan is stand alone or alternatively it can be included in the risk management plan.

While FDA regulation, 21 CFR Part 820 does not require a Usability Engineering Plan/ Human Factors plan, it is useful to outline the overall usability effort, the deliverables and where the activities fall within the overall design and development plan.

Content of Plan:

- Introduction- outlining the product, purpose and scope of the plan
- Roles and Responsibilities to ensure the right stakeholders review and approve
- User Research- user needs assessment, literature review, competitor analysis, task analysis
- User Interface Design- the key characteristics required, critical to quality and other design and usability requirements of the medical device
- Risk Management- Description of how risks are identified and how risks to users is reduced via Usability engineering studies
- Usability Testing- outline of formative and summative evaluations planned

Product Realization Process and Risk Management

Manufacturing companies that are responsible the production, design and development of medical devices are required to have processes and procedures in regards not only to risk management but also Product realization. Risk management and Process realization are normally separate processes with different procedures and SOPs. However, regulations e.g. EU MDR, require that the two processes work together with design and development taking into account risk management. Above all, this is to ensure safety requirements are included in development process and that risk are identified and tracked during the development lifecycle in order to ensure they are addressed or mitigated. The review of the results of the design verification activities during development to verify the risk controls were effective is part of this process.

Formative Studies and Evaluation

Formative studies or evaluation is performed early-on and throughout the develop process using simulations and early working prototypes that explore if general usability principles and intended use can be performed. For medical devices, the main focus of formative studies is on providing a preliminary analyses of the user interface to identify the use safety aspects of user interaction with the device.

Therefore, by completing formative studies early in the product life-cycle, they can inform and help develop suitable user interfaces and identify if any design changes are required. It also serves to uncover any unanticipated use errors. Completing user interface testing via formative evaluations improves the likelihood that the final (summative) evaluation of the usability of the user interface can be completed successfully. Any design changes or modifications to the user interface required should be subject to design control and principles of change management. This is to ensure that changes are reviewed and approved by the appropriate functions and that continuity is maintained. Further evaluations are not required if the specified quality level has been achieved that gives the confidence that the final acceptance criteria will be met when the summative evaluation is conducted at design transfer.

Formative evaluations can include, but are not limited to:

- Hands-on evaluations
- Cognitive Walk-Through
- Customer Surveys
- Focus Groups

The effectiveness of the user interface is challenged by testing the device via formative evaluation. This formative testing can confirm that the user needs and design inputs are fulfilled. It also can identify unknown or unanticipated use errors. The testing done under evaluation may can also assess the effectiveness of risk control measures.

Once a Use error is identified, a root cause should be determined to establish if the use error was a result of an unidentified user need or a weakness in the design of the user interface.

Generally, success depends on completing an evaluation of the user interface in a study with pre-determined acceptance criteria according to the user interface specification. Residual risks related to usability must be controlled to acceptable levels. The manufacturer can apply the acceptance criteria in accordance with their risk policy and ISO 14971.

Design Validation can be demonstrated from a Summative Evaluation report on Human Factors Engineering/ Usability Validation

Formative evaluations can involve simple mock-up devices, preliminary prototypes or more advanced prototypes as the design evolves. They can also be tailored to focus on specific accessories or elements of the user interface or on certain aspects of the use environment or specific sub-groups of users. Design modifications should be implemented and then evaluated for adequacy during this phase of device development in an iterative fashion until the device is ready for human factors validation testing. User interface design flaws identified during formative evaluation can be addressed more easily and less expensively than they could be later in the design process, especially following discovery of design flaws during human factors validation testing. If no formative evaluation is conducted and design flaws are found in the human factors validation testing, then that test essentially becomes a formative evaluation. The effectiveness of formative evaluation for providing better understanding of use issues (and preventing a human factors validation test from becoming a formative evaluation) will depend on the quality of the formative evaluation.

Depending on the rigor of the test you conduct, you might underestimate the existence or importance of problems found, for example, because the test participants were unrealistically well trained, capable, or careful during the test.

Unlike human factors validation testing, company employees can serve as participants in formative evaluation; however, their performance and opinions could be misleading or incomplete if they are not representative of the intended users, are familiar with the device or are hesitant to express their honest opinions. The protocol for a formative evaluation typically specifies the following:

• Evaluation purpose, goals and priorities
• Portion of the user interface to be assessed
• Use scenarios and tasks involved; • Evaluation participants
• Data collection method or methods (e.g., cognitive walk-through, observation, discussion, interview)

A simple kind of formative evaluation involving users is the cognitive walk-through. In a cognitive walk-through, test participants are guided through the process of using a device. During the walk-through, participants are questioned and encouraged to discuss their thought processes (sometimes called "think aloud") and explain any difficulties or concerns they have.

Simulated-Use Testing Simulated-use testing provides a powerful method to study users interacting with the device user interface and performing actual tasks. This kind of testing involves systematic collection of data from test participants using a device, device component or system in realistic use scenarios but under simulated conditions of use (e.g., with the device not powered or used on a manikin rather than an actual patient). In contrast to a cognitive walk-through, simulated-use testing allows participants to use the device more independently and naturally.

Simulated use testing can explore user interaction with the device overall or it can investigate specific human factors considerations identified in the preliminary analyses, such as infrequent or particularly difficult tasks or use scenarios, challenging conditions of use, use by specific user populations, or adequacy of the proposed training.

During formative evaluation, the simulated-use testing methods can be tailored to suit your needs for collecting preliminary data. Data can be obtained by observing participants interacting with the device and interviewing them. Automated data capture can also be used if interactions of interest are subtle, complex, or occur rapidly, making them difficult to observe. The participants can be asked questions or encouraged to "think aloud" while they use the device. They should be interviewed after using the device to obtain their perspectives on device use, particularly related to any use problems that occurred, such as obvious use error. The observation data collection can also include any instances of observed hesitation or apparent confusion, can pause to discuss problems when they arise, or include other data collection methods that might be helpful to inform the design of a specific device user interface.

Formative usability testing is performed early, using simulations and early working prototypes; it is intended to explore whether usability objectives are attainable, but does not have strict acceptance criteria. Types of formative usability tests include the following:

a) **Exploratory testing:** Tests of users performing high-level tasks or walking through the tasks using low-fidelity simulations (e.g., paper sketches of computer screens or crude physical foam models) (see 9.3.4.5). Concepts are tested at this stage of development (e.g., a computer simulation of a touch-screen user interface for a patient monitor or paper sketches of the navigational buttons and menus for such a device).

b) **Comparison (contrast) testing:** Tests comparing two or more design alternatives (e.g., a test to measure the effectiveness and alerting properties of two competing types of auditory alarm signals for an infusion pump).

c) **Comparison (competitive) testing:** Tests that gather usability data related to a competitor's product (e.g., a usability test comparing the task-success rate and time to run a blood chemistry test using a variety of hand-held point-of-care blood analyzers). These tests could be part of the design exploration to understand the best features of existing products.

d) **Assessment testing:** Tests that give users realistic tasks to perform on working prototypes or more fully developed simulations, usually without patients attached to the device (e.g., usability testing of the feel and control of a working prototype of a hand-held glucose meter).

In summary, formative evaluation can reveal previously unrecognized use-related hazards and use errors and help identify new critical tasks. It can also be used to:

- Inform the design of the device user interface (including possible design tradeoffs)
- Assess the effectiveness of measures implemented to reduce or eliminate use-related hazards or potential use errors
- Determine training requirements and inform the design of the labeling and training materials (which should be finalized prior to human factors validation testing)
- Inform the content and structure of the human factors validation testing.

The methods used for formative evaluation should be chosen based on the need for additional understanding and clarification of user interactions with the device user interface. Formative evaluation can be conducted with varying degrees of formality and sample sizes, depending on how much information is needed to inform device design, the complexity of the device and its use, the variability of the user population, or specific conditions of use (e.g., worst-case conditions). Formative evaluations are used to inform device user interface design while it is in development

Efficiency

The concept of efficiency in relation to usability can be understood to be the effectiveness in relation to resources expended. The greater the efficiency with a user often leads to better outcomes and safe use. A lack of efficiency can contribute to risks or increase existing risks. If the medical device has a time based or time related performance characteristic, efficiency may also be of greater importance. An obvious example of a medical device where efficiency of its use is critical to outcomes is Automated External Defibrillators (AEDs).

Usability at Design Intent

Usability is created by characteristics of the user interface that facilitate use, i.e. to make it easier for the user to perceive information presented by the user interface, to understand and to make decisions based on that information, and to interact with the medical device to achieve specified goals in the intended use environments. many of these factors can influence safety to various extents.

Safety

Safety is freedom from unacceptable (use-related) risk.

Whereas, freedom from discomfort device is called 'satisfaction'. The manufacturer must distinguish between safety risks and customer feedback relating to satisfaction of positive experience

Usability Engineering File

The usability engineering file is the collection of documents and records that a manufacturer generated in relation to usability and shows the results of the Usability Engineering Process. The Usability Engineering File allows more efficient auditing.

The Usability Engineering File typically includes:

1) Usability Plan (or integrated into the risk management plan)
2) Use Specification
3) UOUP assessment as applicable
4) Use Related Risk Assessment (URRA)
5) User Interface Specification (can be part of the IOV)
6) Usability Report (can be part of the risk management report)
7) UOUP assessment as applicable
8) Formative Evaluations
9) Usability Engineering Report
10) Post Marketing Surveillance

Updates, as appropriate, should be considered throughout the Usability Engineering Process and monitored through the product lifecycle.

Use Scenario

Use scenarios describe the user interaction with the medical device that aims to achieve a certain result under specific conditions of use. Use scenarios can be written as statements, in story like text or by means of simple bullet points that mimic steps or tasks. Different situations can present different use scenarios that can include correct use or normal use with use error in various use scenarios. A hazardous situation is any circumstance in which people, property or the environment is/are exposed to one or more hazards. When a particular use scenario leads to a hazardous situation, the use scenario is called a hazard-related use scenario. An example of a Use scenario would be going swimming while still wearing a 24hr Blood Pressure monitor.

Environmental Factors and Usability

The use environment of a medical device should be considered in the design planning stage of a product and should result in design inputs. A home setting is distinct from a hospital setting. More specifically, various environments co-exist in hospitals but exhibit different environmental conditions that can result in poor use or use errors. Temperature, humidity, noise, limited lighting, confined spaces can all influence the manner the users ability to perform critical tasks.

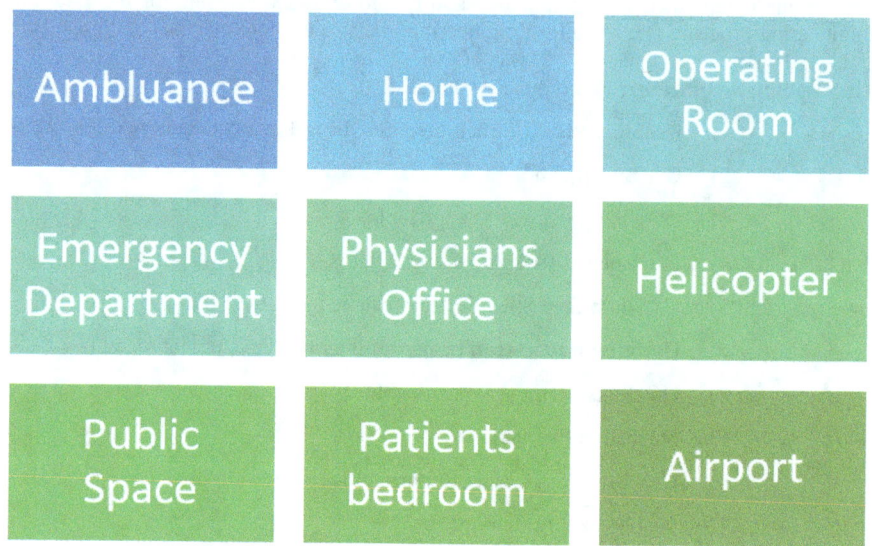

Use Specification

The use specification is a document that provides a summary of the important characteristics related to the context of use of the medical device. The intended medical indication, patient population, part of the body or type of tissue interacted with, user profile, use environment, and operating principle are typical elements of the use specification.

- descriptions of intended device users, uses, use environments, and training
- intended user population(s) and meaningful differences in capabilities between multiple user populations that could affect user interactions with the device
- intended use and operational contexts of use

- use environments and conditions that could affect user interactions with the device
- training intended for users
- documentation required for users

Examples of Intended Users:
- *Laypersons (patients, lay caregivers)*
- *Nurses*
- *Pharmacists*
- *Doctors*

The Intended patient population must also be considered when developing a Use Specification. This may include the intended age group, health, or condition of the patient population. The intended patient population can be similarly referred to as the user profile which also describes: While the above factors (age, health, condition) are the most important attributes, other information such as Occupation, education level, Linguistic and cultural background and potential disabilities can be considered.

Use Environment

Actual conditions and setting in which users interact with the medical device. Factors that should be considered in describing the use environment include, but are not limited to:
- *Lighting*
- *Sound*
- *Sterile or non-sterile, single-use or reusable*
- *Hospital use or home use*
- *Ward or Operating theatre*
- *Ambulance use,*
- *Transportation*
- *Storage*
- *Disposal*
- *Available personal protective equipment*

User Interface

The user interface encompasses the means of interaction between the medical device (including all of the elements) and the user, either by means of software or hardware interfaces. Documentation such as the instructions for use are considered part of the medical device and its user interface. The terms usability engineering, human factors engineering or 'human factors engineering' can be used. However, applying knowledge of people to the user interface design is better described as human factors engineering. While the activity of evaluating devices and their interfaces may better describe usability engineering.

- *Documentation provided*
- *cables*
- *tubing connections*
- *handles*
- *force required to move the weight*
- *work surface height*
- *markings (labelling)*
- *video display (size)*
- *push buttons*
- *touch screens*
- *auditory signals*
- *vibratory signals*
- *visual signals*
- *keyboard and mouse*
- *haptic controls (knobs, joysticks)*

Guidance on User Interface	
Description of User Interface Specification	The medical device be described in straight forward language and account of any associated connections, parts or assemblies that is required during the operation of the medical device by the user
Example	*For an Upper arm Blood pressure monitor, the user interface consists of the unit which houses the batteries, pump, measurement device and digital display. An upper arm cuff connected to a tube and easy-fit connection*
Factors to consider when developing a User Interface Specification	Use Specification All Known or foreseeable use errors associated with the medical device Hazard-related use scenarios
Identify User Needs	*Identify User Needs* The User Needs are written in terms of device attributes and not user tasks. The intent of the user needs section is to help identify points of user interaction with the device. Should a change to the design be made, the impact on the user interaction will be better understood. 1. Does the medical device design allow easy handling 2. If the device is sealed within a sterile barrier, can it be opened aseptically by the user. 3. Does the packaging provide protection during storage and transportation 4. If applicable, sterility of device is maintained during shipping, storage and over course of shelf life.

| | 5. Information printed on the packaging must be locatable and legible by a user with normal eyesight |
| | 6..The device can be removed and assembled easily from the packaging |

User Interfaces of Unknown Provenance (UOUP)

Prior February 2015 products released to the market without any User Interface design changes, are known as User Interfaces of Unknown Provenance (UOUP) per the usability standard- IEC 62366-1 Medical Devices, Application of usability engineering to medical devices. In place of Usability studies and user interface evaluation, a review of Post-Production data can be completed. However, subsequent design changes that impact the usability interface post February 2015 should be evaluated.

User Interface Specification

The user interface specification is made up of the design requirements for a the medical device that document and detail the characteristics of the user interface. User Interface requirements should be S-M-A-R-T. Specific, measurable, achievable, relevant and time bound.

User Needs and Requirements

The User Needs can be generated from various sources including focus groups, existing products, and complaint data.

From the User Needs, requirements are generated to meet these needs which have measurable (testable) acceptance criteria.

These requirements can include documentation that is required for safe use of the product which can include Instructions for Use or a User Manual, as well as training that is required. If training is needed who will conduct the training and what training material will be used is to be defined.

Requirement No.	User Interface Requirement
1	Digital display shall be visible at a distance of 1m to three people standing side-by-side, with all able to read the text on the display screen
2	The medical device shall be capable of producing an auditory alarm of 45dba when measured at 1m from the front of the display screen
3	The medical device shall be mobile and have a wheel locking system for when the unit is stationary
4	The medical device shall be compatible with an 230V electrical supply with integrated surge protection via a fused mechanism
5	The operation of the HMI shall support use of a mouse or keyboard to select commands.

The user interface specification is a source of design inputs for the product and should be subject to design control during the development stages.

The User Interface Specification is part of the development of Design Inputs for the medical device and can be captured in the Design (Inputs/Outputs Verification and Validation Matrix (DIOV).

User Interface Evaluation Plan

The User Interface evaluation plan shall identify the objectives and methods of the required formative and summative evaluations. The plan can serve to detail the user interface evaluation activities and other development activities. For example, the Summative Evaluation is completed using product that is representative of the commercial product and therefore, they must be available when the evaluation is executed. In addition, labelling must be representative in order to create simulated conditions.

Summative Testing

Summative Evaluation is used to confirm the safety of the User Interface while also assessing the effectiveness of risk management measures such as risk controls and mitigations. Summative testing form part of the design validation activities in the development stages of a medical device. By definition, and in keeping with the design control process, design validation is usually in the later stages of development.

In comparison to formative testing, summative testing should apply pre-defined acceptance criteria and account for statistical sample sizes and

Summative Evaluation Protocol

- The summative evaluation demonstrates that the intended users of a medical device can safely and effectively perform critical tasks for the intended uses in the expected use environments.
- All aspects of intended use shall be evaluated.
- For the Summative Evaluation the following is to be used: a production version of the device, representative device users, and actual use or simulated use environment.
- The evaluation can be carried out under conditions of simulated use, but, if possible and necessary, it can be undertaken under conditions of actual use in a clinical study.
- If a clinical study is used the conditions must be the same as an actual use including all instructions for use and training.
- Summative evaluation of usability has formal acceptance criteria.
- Documenting the criteria for determining whether the user has successfully completed the tasks associated with the hazard-related use scenarios is required.
- One possible way to express these criteria is that no use error that leads to a use related harm. Another way is that no use errors lead to unacceptable risk of harm.

Tasks and Use Scenarios

The test protocol should describe the user tasks and/or use scenarios containing tasks to be included in the test, information regarding task criticality or relative priority, and the process by which task inclusion and priority were determined.

The test protocol should also provide a rationale for the extent of device use and the number of times that participants will attempt to use the device.

The Test Participants must represent the population of intended users. If the device has more than one population of users, then the validation testing should be designed to evaluate each distinct user population.

For devices intended to be used by more than one group of users that have distinct abilities or use roles, the number of participants is to be determined based on content and complexity of the device being evaluated. It is recommended that at least 15* participants from each user group should participate in the validation testing.

*15 is the number referenced in FDA's Guidance on Applying Human Factors and Usability Engineering to Medical Devices. The training provided to test participants should represent the training that actual users will receive. Retention of training decays over time, therefore, prior to testing a period should elapse following training. The FDA recommends this should not be less than one hour but should represent the actual decay period between training and first use.

The summative evaluation report documents the results of the Summative Evaluation Plan and Protocol.

The root causes of problems identified during validation testing should be evaluated from the perspective of the test participants involved and direct performance data will support this determination. Data analysis should include subjective feedback from participants regarding critical task experience, difficulties, "close calls," and any task failures by test participants.

Failures and difficulties associated with greater than minimal risk and attributable to the user interface should be addressed by designing and implementing risk mitigation strategies and re-testing those elements to confirm their success at reducing risks to acceptable levels without introducing any new risks. Depending on the level of mitigation strategies required, revalidation may be necessary.

Residual Risk

Residual risk is risk that remains that cannot be eliminated or mitigated through any modifications to the product via design, user interface, accessories, labeling or training.

The analysis of residual risk should determine if the residual risk is outweighed by the advantages offered by the device. If design flaws that could harm the patient or user are identified, planning to address them in subsequent versions of the device post launch is not acceptable.

Post-Marketing Surveillance

Post-Production Surveillance (Post-Market Surveillance) consists of customer feedback and complaint data. Customer feedback may simply be due to dissatisfaction while using the device. More serious complaints that have a patient safety impact often require reporting to regulatory authorities. These are known as reportable events or adverse events depending on the region or competent authority. In addition, for a manufacturer to be compliant ISO 14971, post market surveillance data must be reviewed throughout the lifecycle of the medical device.

In support of post-production activities, the manufacturer must establish, document and maintain a system to actively collect and review information relevant to the medical device in the both the production and post-production phases. The methods of data collection and processing should also be included in the product risk management plan.

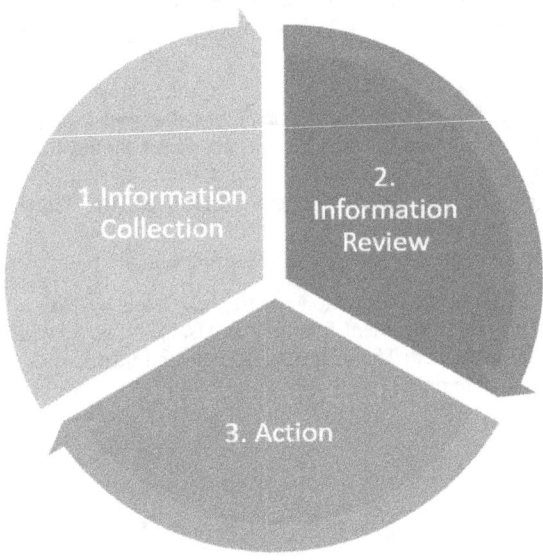

Information collection

The manufacturer must collect, where applicable:

- information generated during production
- information generated by the user (complaints
- information generated by the supply chain;
- publicly available information
- information related to the generally acknowledged state of the art.

Information review

The manufacturer shall review the information collected for possible relevance to safety, especially whether:

- previously unrecognised hazards or hazardous situations are present
- an estimated risk arising from a hazardous situation is no longer acceptable

- the overall residual risk is no longer acceptable in relation to the benefits of the intended use
- the generally acknowledged state of the art has changed. The results of the review shall be recorded in the risk management file

Actions

If the collected information is determined to be relevant to safety, the following actions apply.

Reassessment of risks

- the manufacturer shall review the risk management file and determine if reassessment of risks and/or assessment of new risks is necessary
- Are New Risks or emerging risks identified?

Residual risk

- if a residual risk is no longer acceptable, the impact on previously implemented risk control measures shall be evaluated and should be considered as an input for modification of the medical device;
- the manufacturer should consider the need for actions regarding medical devices on the market
- any decisions and actions shall be recorded in the risk management file.

The collection of data post-production is also mandated by IEC 62366-1. Although all known Use Errors must be identified in the Usability Evaluation, unanticipated Use Errors may not be identified. The information collected post launch can therefore identify any use errors that were missed.

European Regulations- Usability and MDR

GSPR CHAPTER 1, Section 5

In eliminating or reducing risks related to use error, the manufacturer shall:

(a) reduce as far as possible the risks related to the ergonomic features of the device and the environment in which the device is intended to be used (design for patient safety), and

There are several key words and corresponding requirements in section 5 (a). Firstly, there is a requirement to 'reduce as far as possible'. Historically, the term 'ALARP', as low as reasonably possible' was often included in risk management. This is no longer appropriate, and the risk measure must be as low as possible (reasonably has been dropped).

Ergonomic features of the device can be understood to the user interface and human factors engineering. Use scenarios documented by the manufacturer must also account for the environment in which the device is intended to be used.

(b) give consideration to the technical knowledge, experience, education, training and use environment, where applicable, and the medical and physical conditions of intended users (design for lay, professional, disabled or other users).

Part (b) is also intended to eliminate or reduce the risk of use errors by designing products, labelling and instructions for use appropriate user in mind such as a lay person, physician or other.

Per EU MDR 745/2017 a definition of device deficiency includes the term 'user errors' due to an inadequacy.

Article 2, Definitions

(59) 'device deficiency' means any inadequacy in the identity, quality, durability, reliability, safety or performance of an investigational device, including malfunction, <u>use errors</u> or inadequacy in information supplied by the manufacturer;

It is not feasible for a manufacturer to wait until a product is launched to identify use errors. The manufacturer must apply usability engineering to develop use specifications, user interfaces and performance formative and summative evaluations as required. Therefore, the Usability engineering process should identify foreseeable misuses and use errors and via redesign or introduction of protective measures reduce risks to as low as possible, where the benefit outweighs the risks and the risk is acceptable.

Design Controls and Usability Engineering

The FDA Quality system regulation, 21 CFR 820.3 provide manufacturers with Design controls applicable to medical devices. Usability Engineering (Human Factors Engineering, HFE) play an important role in the product development process and the safe design of products.

Usability engineering should begin within the product development process and continue.

Design Control per FDA 21 CFR

'(a) General. (1) Each manufacturer of any class III or class II device, and the class I devices listed in paragraph (a)(2) of this section, shall establish and maintain procedures to control the design of the device in order to ensure that specified design requirements are met.

(2) The following class I devices are subject to design controls:

(i) Devices automated with computer software; and

(ii) The devices listed in the following chart.

Section	Device
868.6810	Catheter, Tracheobronchial Suction.
878.4460	Glove, Surgeon's.

Design and Development Planning

(b) Design and development planning. Each manufacturer shall establish and maintain plans that describe or reference the design and development activities and define responsibility for implementation. The plans shall identify and describe the interfaces with different groups or activities that provide, or result in, input to the design and development process. The plans shall be reviewed, updated, and approved as design and development evolves.

At the planning stage of a project, the user needs should start to be documented in order to create design inputs. Depending on the severity of harm a device can cause can inform the scope and time of the usability assessment. Therefore, the most important factors to consider for Usability engineering at the D&D Planning stage are user needs (Use specification, User interface), identifying harms and gaining a sense of the overall effort.

Design Input

(c) **Design input.** Each manufacturer shall establish and maintain procedures to ensure that the design requirements relating to a device are appropriate and address the intended use of the device, including the needs of the user and patient. The procedures shall include a mechanism for addressing incomplete, ambiguous, or conflicting requirements. The design input requirements shall be documented and shall be reviewed and approved by a designated individual(s). The approval, including the date and signature of the individual(s) approving the requirements, shall be documented.

Usability engineering/ human factors evaluations should be used to define the inputs that relate to user needs. This may be the needs of a patient and/or the user of the medical device.

Design inputs cover not only user needs but factors such as sterility requirements (if applicable), requirements of standards, requirements of local regulation and design for manufacturability. Design controls are usually followed after the first deliverable that will be used to demonstrate compliance to 21 CFR Part 830.30 such as a design and development plan or a requirements document such as a Design inputs outputs verification matrix, (DIOV).

Use-related risk control measures are also design inputs and should be documented through the development of the product

Design Output

(d) *Design output.* Each manufacturer shall establish and maintain procedures for defining and documenting design output in terms that allow an adequate evaluation of conformance to design input requirements. Design output procedures shall contain or make reference to acceptance criteria and shall ensure that those design outputs that are essential for the proper functioning of the device are identified. Design output shall be documented, reviewed, and approved before release. The approval, including the date and signature of the individual(s) approving the output, shall be documented.

Design outputs must be documented and are typically approved by the design teams and stakeholders. Design outputs normally describe an element of the design and provides requirements for the implementation. For example, the physical attributes of a product would be normally detailed in engineering drawings containing measurement values and tolerances. The Engineering drawing is a design output. Design outputs relate back to the design inputs and form evidence that the design input is progressing through the product development process.

Design outputs relating to Usability Engineering include Use-related Risk Analysis' and User Interface specifications.

Design Review

(e) *Design review.* Each manufacturer shall establish and maintain procedures to ensure that formal documented reviews of the design results are planned and conducted at appropriate stages of the device's design development. The procedures shall ensure that participants at each design review include representatives of all functions concerned with the design stage being reviewed and an individual(s) who does not have direct responsibility for the design stage being reviewed, as well as any specialists needed. The results of a design review, including identification of the design, the date, and the individual(s) performing the review, shall be documented in the design history file (the DHF).

Design reviews are required during the product development process and should occur on a routine basis. Design reviews should include the core team involved in the development of the product and stakeholders. The basis of each design review should include a review of activities against the design and development plan, review of open actions from the previous review and ensure that new requirements or information is assessed and appropriately tracked and added to the project plan.

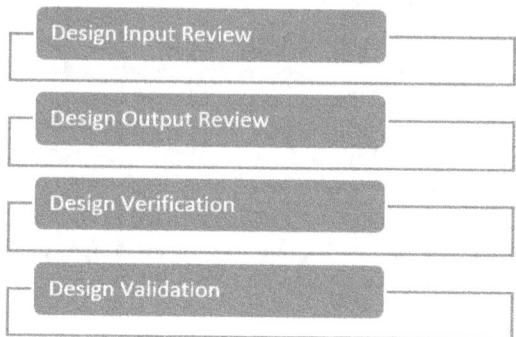

Design Input Review:

- Determine if the design inputs adequately address the user needs and intended use of the medical device?
- Status of the Usability Engineering Plan and are any revisions required?
- Review the actions to be completed in advance of the next design review; e.g. formative evaluation, URRA update.
- Develop user requirements testable for summative testing
- Include Design outputs from formative evaluations in the DIOV.

Design Output Review

- Review results of formative evaluations for use-errors
- Input use errors into risk assessments
- Review any design changes that may impact design outputs
- Review the actions to be completed in advance of the next design review; e.g. formative evaluation, URRA update

Design Verification Review

- Confirm completion of Design verifications
- Verify use related risk controls are implemented or tracked

Design Validation Review

- Summative Evaluations have been completed
- Usability related Risk Control measures are effective
- The benefits outweigh the risks and residual risk is acceptable

(f) *Design verification.* Each manufacturer shall establish and maintain procedures for verifying the device design. Design verification shall confirm that the design output meets the design input requirements. The results of the design verification, including identification of the design, method(s), the date, and the individual(s) performing the verification, shall be documented in the DHF.

Design verification include stability studies, process validations and other studies that confirm design inputs are successfully achieved according to the design output

(g) *Design validation.* Each manufacturer shall establish and maintain procedures for validating the device design. Design validation shall be performed under defined operating conditions on initial production units, lots, or batches, or their equivalents. Design validation shall ensure that devices conform to defined user needs and intended uses and shall include testing of production units under actual or simulated use conditions. Design validation shall include software validation and risk analysis, where appropriate. The results of the design validation, including identification of the design, method(s), the date, and the individual(s) performing the validation, shall be documented in the DHF.

Design validation involves the completion of studies such as usability evaluations and clinical studies that demonstrate the safe and effective use of a device. The conditions of testing should represent actual or simulated conditions that represent realistic use. The participants of the study should include users within the target population. At this point the development stage, the remaining risk of use-errors should be low, and if present still have appropriate risk controls that are effective.

(h) *Design transfer.* Each manufacturer shall establish and maintain procedures to ensure that the device design is correctly translated into production specifications.

Design transfer occurs when the device design is required to be translated into a validation manufacturing process for the purposes of commercial manufacturing. To facilitate the production process, design specification, manufacturing specification and finished product specifications are required to be approved and validated to enable commercial manufacturing. At this late stage in the process the Usability Engineering /Human Factors effort should be largely completed with evaluations reported and actions and risk controls validated and effective.

(i) *Design changes.* Each manufacturer shall establish and maintain procedures for the identification, documentation, validation or where appropriate verification, review, and approval of design changes before their implementation.

Design changes result in changes to design inputs or design outputs. Design changes include modification to product specifications, material changes, changes to product requirements, changes to product features or functionality. Changes to manufacturing processes can also impact the finished product and as such should be assessed for the impact on the design and functionality of the product.

Therefore, any changes that may impact the usability of a medical device should be treated as a design change and must follow a process of impact assessment/evaluation of the design change, design reverification and revalidation and risk re-evaluation. Evaluation of the design change should ask:

- Does the proposed change impact users?
- Is there an impact on safety?
- Is there an impact on effectiveness?
- Is there an impact on Usability?
- Is there an impact on design verifications or validations?

(j) *Design history file*. Each manufacturer shall establish and maintain a DHF for each type of device. The DHF shall contain or reference the records necessary to demonstrate that the design was developed in accordance with the approved design plan and the requirements of this part.

From a usability engineering perspective, the evaluations form part of the verifications and validations associated with the device design. Therefore, relevant studies should be included in the design history file.

Appendix- 4- Simple format of a Use Related Risk Analysis

Use Related Risk Analysis- Upper Arm Blood Pressure Monitor, Rev 1.0

User Task	Identify Potential Use Error	Hazard, Harm	Severity	Risk Control Measures	Risk Control Effectiveness Y/N
1. Connecting the air house	Plug on air house not firmly pushed into position	BP reading not possible, ERR_CUFF	1, Inconvenience	Inherent by Device Design- Plug is tapered to provide easy compliance Labelling/Safety information- Instructions provide a labelled diagram IFUXX00X: Precautions and Warnings	Y
2. Selecting the correct cuff	Selects wrong cuff size	Cuff size too large or too small leading to inaccurate readings	4, lead to undiagnosed hypertension and/or hypotension	Inherent by Device Design- Arm circumference design meets 80% of patient population Labelling/Safety information- Arm size is printed on each cuff Instructions provide step by step guidance and diagram of proper fit range IFUXX00X: Precautions and Warnings	Y
3. Applying the arm cuff	Does not remove clothing covering arm Does not position and orientate the cuff correctly	Prevents accurate readings	4, lead to undiagnosed hypertension and/or hypotension	Inherent by Device Design- Diagram is printed on the cuff Labelling/Safety information- instructions on application IFUXX00X: Precautions and Warnings	Y
4. Start measureme	Removes cuff during measurement	Prevents accurate	4, lead to undiagnosed hypertensio	Inherent by Device Design-inflation of cuff completes in 10seconds Labelling/Safety information- Position and behaviour	Y

					during measurement detailed in the instructions IFUXX00X: Precautions and Warnings	
5. Cuff removal	Starts measurement cycle inadvertently	Compression of arm	2, low to moderate discomfort	Inherent by Device Design-Cuff flap can be removed easily Labelling/Safety information- guidance on remeasurement specified in the instructions for use IFUXX00X: Precautions and Warnings	Y

Severity Scoring and Descriptions

Severity Score	Description
5	Patient death
4	Permanent harm, if condition left unresolved
3	Moderate injury, no lasting effects
2	Moderate discomfort, no lasting effects
1	Discomfort, transient
0	No harm to user or patient

Information Supplied and Usability

Information for safety is a risk control measure that should be used only after the manufacturer has determined that (further) risk reduction by other measures is not practicable.

Information provided that relates to safety must be assessed through the usability engineering process in order to determine the information is:

- perceivable by users in use environments
- is understandable by users
- allows correct use of the medical device

Risk reduction achieved using modified design features that make the medical device inherently safe is always the preferred course of action. If this is not possible, implementing protective measures can be applied.

Information for safety is instructive and gives the user clear instructions of what actions to take or to avoid, in order to prevent a hazardous situation or harm from occurring. This information can be provided in the form of:
- warnings
- precautions
- contra-indications
- instructions for use
- training

The purpose of Warnings and precautions is to identify adverse reactions and potential safety hazards tat are serious or clinically significant. For adverse reactions to be included in patient information there should be causal association with the product and the adverse event.

A serious adverse reaction resulting in one of the following and should be considered for inclusion in the Warnings and Precautions section of patient literature.
- Death
- Life altering adverse event- significant incapacity
- Hospitalization
- Congenital anomaly or birth defect

Contra-indications should be indicated where the clinical situations results in a risk that outweighs the benefits of the device.

Instructions for use

When developing information for safety, it is important to identify to whom this information is to be provided and how it is to be provided. This can include an explanation of the risk, the consequences of exposure and what should be done or avoided to prevent any harm. The manufacturer should consider:

- regulatory requirement
- the need to classify the information for safety, based on the level of risk;
- the level of detail necessary to convey the information for safety;
- the location for the information for safety (e.g. a warning label on the medical device);
- the wording, pictures or symbols to be used to ensure clarity and understandability;
- the intended recipients (e.g. users, service personnel, installers, patients); the appropriate media for providing the information, (e.g. instructions for use, labels, warnings in
- the user interface);

Information for safety can be communicated in different ways, depending on when in the medical device life cycle the information is to be communicated, e.g. via the user interface of a menu-driven medical device, as cautionary statements in the accompanying documentation, or in an advisory notice.

Information for safety can be given in various forms, such as warning labels attached to the medical device, warning statements in the instructions for use, instructions on a graphical user interface, or instructions in training videos.

- Warning: Do not step on surface
- Warning do not freeze
- Warning Do not refrigerate
- Warning: Do not use if seal is broken
- Warning: Do not remove cover, risk of electric shock

Label Design and User Interaction

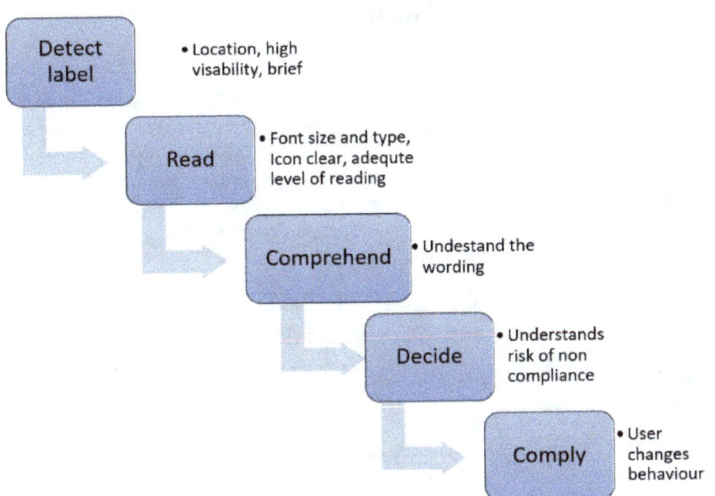

Risk Management Plan

Notes to Author

1. *Blue italic text is for general guidance purposes. Delete for document approval*
2. *Black italic text is text can should be edited for specific risk management plans, as required*
3. *When document is created locally, pagination should be applied*
4. *An approved template should form the basis of a risk management plan. The template should be revision controlled and each page should identify the document title and document ID.*

Document Title:	Risk Management Plan for *Advance 101, Battery Powered Digital Blood Pressure Monitor, UA1101*
Document ID:	R7100-21
Revision:	A
Issued	30 Feb 2022

Notes of Scope of Activities

1. *Risk management plans must cover the full lifecycle of the product, starting with product development to post launch product lifecycle risk management.*
2. *The extent of planned activities and the level of detail of the risk management plan should be commensurate with the level of risk associated with the medical device.*
3. *The requirements in ISO 14971:2019 are the minimum requirements for a risk management plan. Manufacturers can include other items such as time-schedule, risk analysis tools, or a rationale for the choice of specific risk acceptability criteria.*
4. *Risk management plan can also applies to the product realization process (design, development and production of the medical device).*
5. *Other elements can apply to the production and post-production phase (such as installation, use, maintenance, decommissioning and disposal of the medical device).*

| 1.0 Scope of Activities

ISO 14971, 4.4, a) | This risk management plan applies to the risk management activities, the responsibilities and authorities of those involved, the criteria for risk acceptability, the production and post-production information to be collected and reviewed for *Advance 101, Battery Powered Digital Blood Pressure Monitor, UA1101* , and all risk management activities that are carried out during the entire product life cycle.

This risk management plan will be reviewed and updated throughout the product life cycle as new information becomes |

	available.

2.0 Device Description *ISO 14971, 4.4, a)*	Advance 101, UA1101, is a battery powered digital blood pressure monitor, for the measurement of blood pressure via upper arm constriction using a pressurized cuff. The device is intended for use in a home healthcare environment
2.1 Product Names	The device is sold under the following trade names: • Advance 101, Battery Powered Digital Blood Pressure Monitor, UA1101 (US & European Market) • Advantus 101, Battery Powered Digital Blood Pressure Monitor, UAL1101 (Latin America)
2.2 Intended Use	The device is designed and manufactured to measure blood pressure and pulse rate of people for diagnosis. It is intended for use on adults only. The device is suitable for home healthcare and is to be used in It is recommended that blood pressure monitoring is conducted while liaison with a qualified physician.

The functions identified below are responsible for the review and approval of this Risk Management Plan.

Notes on Responsibilities and Authorities

1. *Reviewers and approvers of the Risk management plan must be competent and knowledgeable. Training to Risk management procedures is fundamental.*
2. *ISO 14971 does not specify the functions required. This is the responsibility of the manufacturer and should be based on the nature of the device and the risk management procedures.*

3.0 Responsibilities and Authorities ISO 14971, 4.4 a)	Function	
	Function	**Technical Expertise**
	Risk Management SME	Knowledge of the risk management process and ISO 14971: application of risk management for medical devices and appropriate regulations
	Device R&D	Provides technical knowledge on the operating characteristics and performance of the device
	Engineering	Supports the risks management process with process and manufacturing knowledge and experience
	Operations	Provides Knowledge of the manufacturing process
	Quality	Responsible for the consistent application of procedures
	Clinical Affairs	Provide expertise and clinical evaluation
	Regulatory Affairs	Reviews for compliance to regulations requirements
	Medical Expert	Provides medical expertise, supports literature reviews and other activities
	Nonclinical	Directs and executes the investigation and reporting of nonclinical testing

Notes on Risk Acceptability
1. For each risk management plan the manufacturer needs to establish risk acceptability criteria that are appropriate for the particular medical device
2. It is important to establish the criteria for risk acceptability before starting the risk assessment. Otherwise, the results of the risk assessment could influence the decision when establishing the criteria.

4.0 Risk Acceptability ISO 14971 4.4	The criteria for risk acceptability is established in the policy for determining acceptable risk. The methodology used to evaluate the overall residual risk, and criteria for acceptability of the overall residual risk based on the policy for determining acceptable risk below.

d)

Risk Management Policy

The Risk Management policy for diagnostic devices is intended to provide safe, reliable and effective products to our customers when the products are used in accordance with specified operating instructions.
Acceptability of risks is defined in the Risk Management Plan. Risks are identified in the risk management documents. All identified safety related risks are mitigated to as low as possible, where residual risks remain, a risk-benefit analysis shall be performed.

Where the probability of occurrence of harm cannot be estimated, the criteria for risk acceptability shall be based on the severity of harm alone.

The evaluation of the overall residual risk is determined upon the review of data and literature for the medical device and similar medical devices on the market which is reviewed by the cross-functional team including medical and clinical expertise.

Probability:

Term	Value	Probability per opportunity	Parts per million opportunities
Frequent	5	>1/100	>10,000
Probable	4	1/1,000 - 1/1,00	1000-10,000
Occasional	3	1/10,000 - 1/1,000	100-1000
Remote	2	1/100,000 - 1/10,000	10-100
Rarely	1	<1/100,000	<10

Severity:

Term (Severity)	Severity Value (S)	Description
Catastrophic	5	• Patient Death • Destruction of Facility
Critical	4	• Permanent Impairment or life threatening injury – blindness • Destruction of a piece of capital equipment
Serious	3	• Injury or impairment requiring professional medical intervention • Failure of equipment requiring postponement to second surgery • Damage to equipment or facility requiring repair by technicians or contractors
Minor	2	• Temporary injury or impairment not requiring professional medical intervention • Damage to equipment or facility requiring repair by users
Negligible	1	• Inconvenience or temporary discomfort • Delay of start of surgery, or interruption of surgery of less than 30 minutes
None	0	• No harm to patient or user • Delay of start of surgery, or interruption of surgery of less than 30 minutes • Equipment may not work, but no harm to other equipment or facility

Risk Matrix:

P5 Frequent	Acceptable	Acceptable	Acceptable	Unacceptable	Unacceptable	Unacceptable
P4 Probable	Acceptable	Acceptable	Acceptable	Unacceptable	Unacceptable	Unacceptable

		S0 None	S1 Negligible	S2 Minor	S3 Serious	S4 Critical	S5 Catastrophic
P3 Occasional		Acceptable	Acceptable	Acceptable	Acceptable	Unacceptable	Unacceptable
P2 Remote		Acceptable	Acceptable	Acceptable	Acceptable	Unacceptable	Unacceptable
P1 Rarely		Acceptable	Acceptable	Acceptable	Acceptable	Acceptable	Unacceptable

ISO/TR 24971

- Risks identified shall be assessed for acceptability based on the application of risk estimation and risk analysis
- All residual risks must meet the acceptable residual risk determination criteria. The overall residual risk will be addressed in the risk management report.
- If device data is not available on the probability of occurrence of harm, the acceptance of a risk shall be on the basis of the nature and severity of the harm.
- Determination of acceptable risk for the device is based on applicable standards, comparison of risk from medical devices already on the market and evaluation of clinical data.

Notes on Verification of Implementation

1. The risk management plan is required to specifies how the two verification activities required completed.
2. Verification of implementation of risk control measures can be part of design review, approval of specifications, design and development verification in a quality management system, or other verification activities in a quality management system.
3. Verification of the effectiveness of risk control measures can be part of design and development verification in a quality management system. It can require the collection of clinical data, usability studies, etc., as part of design and development validation in a quality management system.
4. FMEAs should be developed, reviewed and approved for each product based on company procedures.

5.0 Verification of Implementation	The verification of implementation of risk control measures are documented in the Failure Modes Effects and Analysis. These include:

	• Design Failure modes & Effects Analysis (DFMEA) • Process Failure Modes & Effects Analysis (PFMEA) • Use Related Failure Modes & Effects Analysis (UFMEA) • Design Risk Analysis • **Risk Identification:** Potential risks are recorded in each FMEA, based on review of the data available, information from similar devices and information from design verification and design validation activities, • **Risk Estimation:** Where applicable, a Risk Priority Number (RPN) shall be used to quantitatively provide risk estimation for potential hazards. The RPN shall be calculated from the Severity, Occurrence and Detection scoring per FMEAs. All identified safety risks must be mitigated to as low as possible. If, where, residual risk remains, a benefit-risk analysis shall be completed. • **Risk Control Measures:** FMEAs shall be reviewed and updated throughout design and development, post launch and over the life cycle of the product Risk control and mitigation actions shall maintained throughout the lifecycle. • **Risk Acceptance:** All risk identified shall be reduced to as low as possible with available control measures and considered acceptable. Safety risks that cannot be mitigated to as low as possible must be evaluated against a risk/benefit analysis and must meet the acceptable residual risk determination criteria.
ISO 14971, 4.4, f)	

Notes of Effectiveness Review

1. *It is a requirement for the effectiveness of the risk control measures to be verified.*
2. *The results of this verification shall be recorded in the risk management file such as the Risk Management Report*

6.0 Effectiveness review ISO 14971, 7.2	The effectiveness of the risk control measures shall be verified and documented. Verification of the Risk control measures shall be recorded in the risk management report and is subject to approval by a cross functional group. Verification of effectiveness can also be performed over the course of the life cycle or the medical device to ensure the effectiveness of risk control measures

	remain current and meet the requirements of risk acceptability. The risk mitigation measures in the risk documentation (e.g. PFMEA, UFMEA, DFMEA) shall be reviewed to determine effectiveness. The result of risk mitigation activities shall determine if risks have been reduced as far as possible or if the risk can be reduced further
7.0 Data Collection *ISO 14971, 4.4 g)*	Production and Post production data should take the following sources into account: • Design Changes and Change Controls • Product Quality Review • Clinical Evaluation Reports • Cross Functional Risk Assessments • Complaint Analysis and trending • Customer Feedback • Management Review • Quality Control Data • Regulatory Feedback and reporting
Production and Post Production Information *ISO 14971, 4.4 g)*	Production and Post-production data shall be made available to the risk management process and cross functional teams responsible for reviewing risk and risk acceptability. Use Errors shall also form part of post-production data trending based on analysis of complaints data. The processes of the Quality management system shall be utilized to provide data and information that is reliable and current. The frequency of review of the collected data and information shall be commensurate with the level of residual risk and severity of risks based on expert review and clinical

ISO 14971:2019 requires that changes to the risk management plan be recorded in the risk management file.

Risk Management References ISO 14971, 4.5	References to the below risk documents shall be maintained in the risk file and shall be listed in the Risk Management Report
	- Risk Management Plan - Design Risk Analysis - Use Related Failure Modes & Effects Analysis (DFMEA) - Process Failure Modes & Effects Analysis (PFMEA) - Risk Management Report

Revision History	Description
A	*Initial version*

Function	Approver Name	Date
Risk Management Representative	*Approvers signature*	*Date of approval*
Device R&D	*Approvers signature*	*Date of approval*
Engineering	*Approvers signature*	*Date of approval*
Operations	*Approvers signature*	*Date of approval*
Quality	*Approvers signature*	*Date of approval*
Clinical Affairs	*Approvers signature*	*Date of approval*
Regulatory Affairs	*Approvers signature*	*Date of approval*
Medical Expert	*Approvers signature*	*Date of approval*
Nonclinical	*Approvers signature*	*Date of approval*

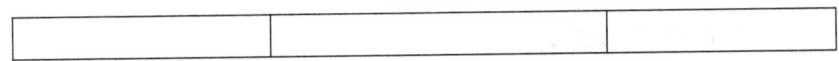

Disclaimer: Template is intended for general learning and general training purposes. While every effort has been made by the author to ensure informative and accurate information is provided, the author and publisher shall not be responsible for any error or omission, and or subsequent material loss, financial, personal, private, public, property or physical damage or harm.

It is the responsibility of individuals to conform with all legal and regulatory requirements in regard to risk management, and to apply the standards, practices and applicable regulations to the applications, products, business and commercial activities they are involved with.

Useful Definitions

Design Verification- Design verification shall confirm that the design output meets the design input requirements.

User- person interacting with (i.e. operating or handling) the medical device.

Layperson-Someone who is not an expert in or does not have a detailed knowledge of a particular subject, in this case the operation or use of the medical device. E.g. the patient or caregiver.

Design Input-a physical and performance requirement of a device that is used as the basis of design.

Design Output-from the results of the design effort and activity, a design output details the device, packaging, manufacturing, testing and labelling requirements

Design Validation- Design Validation is establishing by objective evidence that device specifications conform with user needs and intended use(s).

Input Output Verification Validation Matrix- The traceability matrix created during product development or as part of a design change that defines user needs, inputs, outputs, design verification and design validation.

User Profile-Summary of the mental, physical and demographic traits of an intended user group, as well as any special characteristics, such as occupational skills, job requirements and working conditions, which can have a bearing on design decisions.

User Interface of Unknown Provenance (UOUP)- if the User Interface of a medical device was previously developed (prior to publication of standards) for which adequate records of usability process and Summative Evaluation is not available.

www.ingramcontent.com/pod-product-compliance
Lightning Source LLC
Chambersburg PA
CBHW070310220526
45465CB00004B/1827